Surviving the

EARTHQUAKE

HEAR MY STORY

Linda Barghoorn

CRABTREE
PUBLISHING COMPANY
WWW.CRABTREEBOOKS.COM

Author: Linda Barghoorn

Editorial director: Kathy Middleton

Editors: Sarah Eason, Jennifer Sanderson, and
  Ellen Rodger

Proofreaders: Tracey Kelly, Melissa Boyce

Editorial director: Kathy Middleton

Design: Paul Myerscough

Cover design: Margaret Amy Salter

Photo research: Rachel Blount

**Production coordinator and
  Prepress technician:** Tammy McGarr

Print coordinator: Katherine Berti

Consultant: John Farndon

Produced for Crabtree Publishing Company by Calcium

Photo Credits:
t=Top, c=Center, b=Bottom, l= Left, r=Right

Inside: Shutterstock: Amorat Raj: pp. 12–13, 13r, 23r;
Arindambanerjee: p. 19; Stefano Barzellotti: p. 7r; Fiki
J Bhayangkara: pp. 1, 16–17, 17b, 28–29; Bonandbon:
p. 15b; ChameleonsEye: p. 27; Frans Delian: p. 11;
Everett Historical: p.9; Peter Hermes Furian: pp. 14–15t;
HikoPhotography: p. 26; Agung loningkito: pp. 22–23; Fajrul
Islam: p. 28b; Lakeview Images: p. 18; Lucky Team Studio:
p. 8; Dudarev Mikhail: p. 5; mTaira: p. 21; Amirudin Agus
Nursalim: pp. 6–7; Shigemi Okano: p. 25; Prometheus72: p.
4; Snehit: p. 20; Jason Steel: p. 24; VectorMine: p. 10.

Cover: Shutterstock: Herwin Bahar

Publisher's Note: The story presented in this book is a
fictional account based on extensive research of real-life
accounts, with the aim of reflecting the true experience of
victims of natural disasters.

Library and Archives Canada Cataloguing in Publication

Title: Surviving the earthquake : hear my story / Linda Barghoorn.
Names: Barghoorn, Linda, author.
Description: Series statement: Disaster diaries | Includes index.
Identifiers: Canadiana (print) 20200151347 |
  Canadiana (ebook) 20200151355 |
  ISBN 9780778769880 (hardcover) |
  ISBN 9780778771142 (softcover) |
  ISBN 9781427124456 (HTML)
Subjects: LCSH: Earthquakes—Indonesia—Sulawesi—
  Juvenile literature. | LCSH: Earthquakes—Juvenile literature. |
  LCSH: Disaster victims—Juvenile literature.
Classification: LCC QE537.2.I5 B37 2020 | DDC j551.2209598/4—dc23

Library of Congress Cataloging-in-Publication Data

Names: Barghoorn, Linda, author.
Title: Surviving the earthquake : hear my story / Linda Barghoorn.
Description: New York : Crabtree Publishing Company, [2020] |
  Series: Disaster diaries | Includes index.
Identifiers: LCCN 2019057508 (print) | LCCN 2019057509 (ebook) |
  ISBN 9780778769880 (hardcover) |
  ISBN 9780778771142 (paperback) |
  ISBN 9781427124456 (ebook)
Subjects: LCSH: Earthquakes--Indonesia--Sulawesi--Juvenile
  literature. | Earthquakes--Juvenile literature. | Disaster victims-
  -Juvenile literature. | Indonesia--History--21st century--Juvenile
  literature.
Classification: LCC QE521.3 .B337 2020 (print) |
  LCC QE521.3 (ebook) | DDC 363.34/95092--dc23
LC record available at https://lccn.loc.gov/2019057508
LC ebook record available at https://lccn.loc.gov/2019057509

Crabtree Publishing Company

www.crabtreebooks.com   1-800-387-7650

Printed in the U.S.A./022020/CG20200102

Published in Canada
Crabtree Publishing
616 Welland Ave.
St. Catharines, Ontario
L2M 5V6

Published in the United States
Crabtree Publishing
PMB 59051
350 Fifth Avenue, 59th Floor
New York, New York 10118

Published in the United Kingdom
Crabtree Publishing
Maritime House
Basin Road North, Hove
BN41 1WR

Published in Australia
Crabtree Publishing
3 Charles Street
Coburg North
VIC, 3058

Contents

Earthquakes and Their Victims

Forces deep beneath the surface of Earth have been shaping our planet's landscapes for millions of years. Large pieces of rock continuously move and shift underground. When this rock moves or shifts suddenly, it causes tremors, or shaking, beneath Earth's surface, creating the **natural disasters** we know as earthquakes.

Everyday Occurrences

Earthquakes happen somewhere in the world every day. Often the tremors are so small that we do not even notice them. But stronger earthquakes have the power to cause massive destruction, killing thousands of people and devastating cities.

Collapsing buildings are the single largest cause of death during earthquakes.

Down the Slopes

In mountainous regions, earthquakes often cause huge walls of mud or rock to break off hillsides. Powerful **mudslides** or **landslides** gather speed as they push down the slopes, burying villages and roads, blocking rivers, and flattening forests.

Roads can be damaged, making them impossible to use.

Under the Sea

When an earthquake takes place beneath the ocean's floor, it can create a **tsunami**—a series of large waves that can move at speeds of 225 miles per hour (362 kph). When a tsunami crashes onto land at such high speeds, it sweeps away everything in its path in minutes.

VIOLA'S STORY

In this book, you can find out what it is like to live through a natural disaster by reading the **fictional** story of Viola, a young girl caught up in the 2018 earthquake on the Indonesian island of Sulawesi. Look for her story on pages 6–7, 12–13, 16–17, 22–23, and 28–29.

Restless Gods

In ancient times, people believed that earthquakes were caused by gods punishing Earth. Today, we know that they are part of the natural forces that continue to reshape Earth's landscapes. Whatever their cause, they can bring about death and disaster on an enormous scale.

VIOLA'S STORY:
Before the Quake

My name is Viola and I grew up in Balaroa, a village in the leafy hills of the Indonesian island of Sulawesi. We moved there when I was little because Papa got a government job in the nearby city of Palu. Our village had grown quickly to provide homes for people working in Palu. Our community was a close—knit one, where aunties, uncles, cousins, and friends lived close by, and we saw each other often. I have two older brothers, Adi and Putra, and a younger sister, Intan.

While Palu looks like paradise, it has a dangerous secret. Scientists have always known that earthquakes threaten the peace of this place, but few others paid attention. The last earthquake happened when I was just a baby, in 2005, so I don't remember it. People say the

The city of Palu is located on rich, *fertile* land between a long, curved bay and lush mountains.

Looking back, I can't believe that the beautiful place we lived in was so dangerous.

electricity went out for a short time, and several stores in Palu were damaged. One person was killed, and a few were injured. Afterward, people breathed a sigh of relief that the earthquake was over and continued with their lives. Government officials ignored scientists' warnings about the dangers of building a city like Palu on soft, earthquake–prone ground. Few people who lived there were truly aware of the real danger that lurked underground.

That all changed on September 28, 2018, when a violent earthquake struck Palu. That day had started as a celebration as the city prepared for the important festival of Palu Nomoni. It celebrates the city's history with costumes, parades, food, and family get–togethers. Our little house was bustling with excitement as Mama cooked a festive dinner and we waited for our relatives to arrive. At around 2:30 p.m., the ground beneath us shook slightly, but we were so caught up in the excitement that we barely noticed.

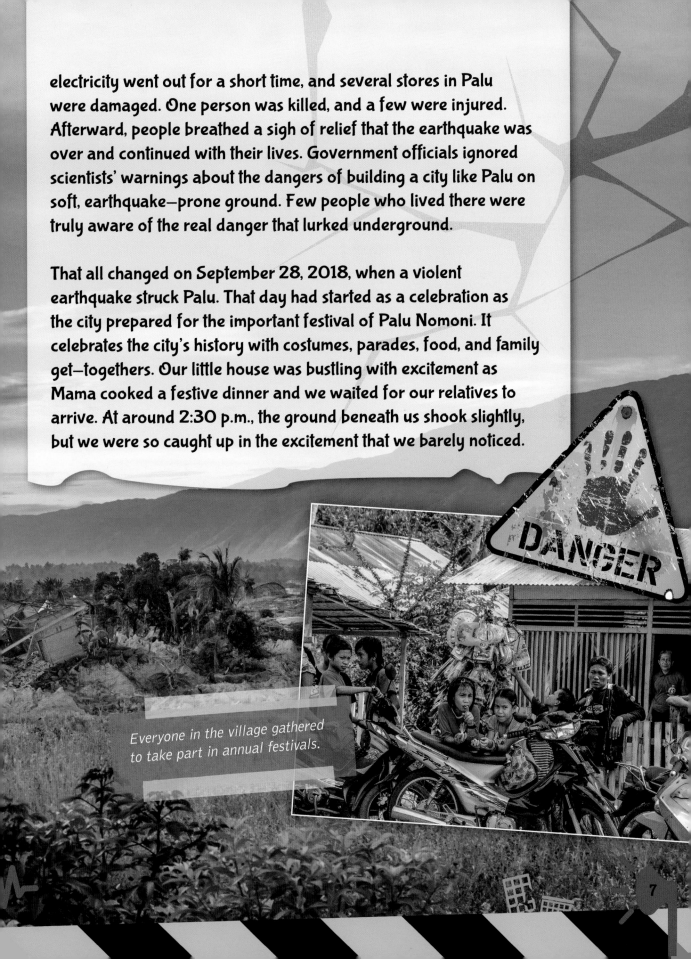

Everyone in the village gathered to take part in annual festivals.

A Trail of Destruction

Scientists estimate that Earth experiences more than 1 million earthquakes every year. Most are so small in **magnitude** that people do not even feel them. These are known as microquakes. Other earthquakes happen in areas that are quite remote—far away from where people live and work—so they go almost completely unnoticed and cause little disruption to people's lives.

Landslides, such as this one in Italy in 2009, can be triggered by earthquakes.

Power Quakes

Sometimes, an earthquake is so ferocious that people are literally shaken up by its extreme **intensity**. Areas that are densely populated—where millions of people live and work—are most vulnerable to the forces of powerful earthquakes.

Throughout history, earthquakes have caused billions of dollars of damage to cities, roads, bridges, homes, pipelines, and power lines. In the last 10 years, more than 650,000 people have been killed worldwide as a result of earthquakes and the tsunamis, **avalanches**, and mudslides that they have triggered.

The Most Powerful

The deadliest earthquake in recorded human history happened more than 450 years ago in the Shanxi province of China. Ancient city walls and temples were totally destroyed. Thousands of houses and government buildings crumbled. More than 800,000 people were killed, which is more than the modern-day population of Washington, D.C.

The strongest earthquake in human history took place in Chile in 1960. It was given a magnitude of 9.5 on a traditional scale that measures **seismic** intensity from 0 to 10. While it caused more than $4 billion in damage, fewer than 2,000 people were killed because the region was not densely populated.

VICTIMS OF THE FIRES

The earth beneath San Francisco is constantly shifting and moving, making it extremely vulnerable to earthquakes. In 1906, an earthquake caused three fires to rage across the city, destroying nearly 500 city blocks. More than half the city's 400,000 residents were left homeless, and survivors stood in long lines for food. Today, better building codes and disaster readiness plans have been developed to reduce the risk of earthquake deaths and destruction.

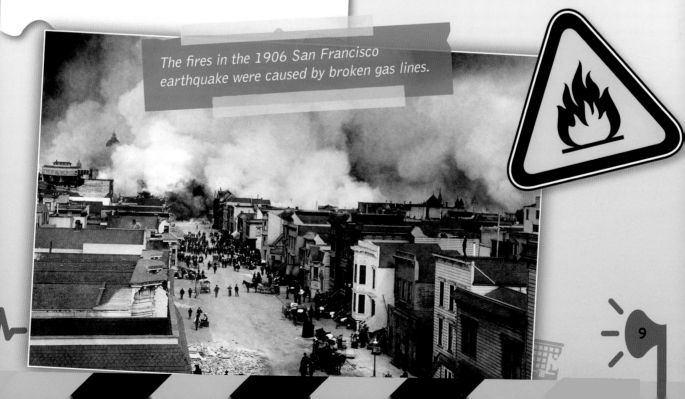

The fires in the 1906 San Francisco earthquake were caused by broken gas lines.

9

How Earthquakes Work

Although we rarely think about it, Earth is constantly shifting beneath our feet. The massive sheets of land, called **tectonic plates**, that make up Earth's surface fit together like puzzle pieces. They are always moving and shifting. Their edges are called plate boundaries. As these plates move toward one another, they can become stuck. This causes stress to build up since the rock is unable to move.

Seismic Waves

When this stress becomes too great, the plates suddenly shift and move. This releases energy in the form of **seismic waves**, which cause the ground to shake. These waves originate, or start, from the focus. The place where these waves reach Earth's surface is called the epicenter. The epicenter is the site of the most violent surface tremors.

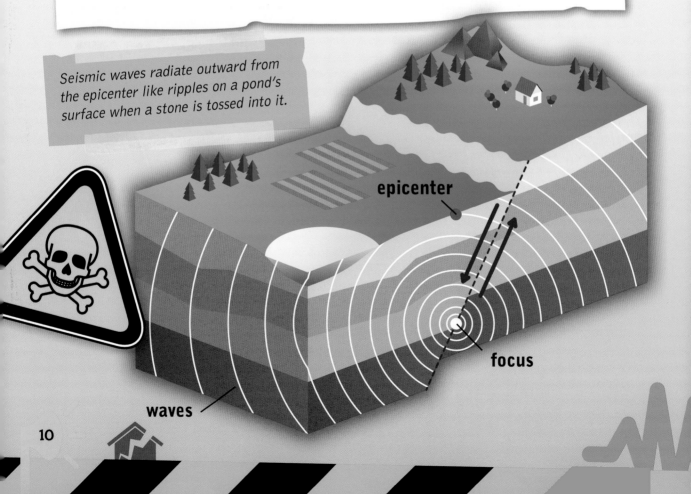

Seismic waves radiate outward from the epicenter like ripples on a pond's surface when a stone is tossed into it.

epicenter

focus

waves

Tremors

When the tremors reach the surface, the ground shakes, creating an earthquake. The main tremor is called the main shock. Sometimes a series of smaller tremors, called foreshocks, come first. The main shocks are often followed by a series of aftershocks. Although they are usually smaller in magnitude, they are dangerous because the main shock will have weakened the structure of buildings, and rescue operations could already be underway, putting the lives of even more people in danger.

Waves more than 100 feet (30 m) high swept away villages, cars, trees, and people in the 2004 Indonesian earthquake (see below left).

VICTIMS OF THE WAVES

In 2004, an earthquake happened off the coast of Indonesia. It triggered a massive tsunami, whose waves traveled 500 miles per hour (805 kph) across the Indian Ocean, reaching more than 12 countries. More than 230,000 people were killed in just a few hours. Scientists claim that the earthquake's enormous 9.1 magnitude tremor was so strong that it wobbled Earth's **axis** by almost 1 inch (2.5 cm).

Measuring Magnitude

An instrument called a **seismograph** measures the seismic waves to determine the earthquake's magnitude. The Moment Magnitude Scale (MMS) is then used to show the magnitude. The greater the magnitude, the more destructive and deadly an earthquake is likely to be.

VIOLA'S STORY:
Earthquake Strike

DANGER

Everyone gets together when there is a celebration. Soon, our aunties arrived carrying plates of food. Papa and my uncles brought our Nana from her home down the street. Everyone chatted excitedly as the delicious meal that had been prepared by Mama and my aunties was laid out on the table. We had just sat down together when we heard a noise far off in the distance. A low, steady rumbling grew louder and closer. It sounded like long rolls of thunder creeping closer toward our village.

Then the ground began to shake and the house began to creak and sway. Dishes slid off the table and crashed to the floor, shattering into pieces as the rumbling sound became a deafening roar. The tiles on the floor began to crack, and dust fell from the ceiling. Everyone jumped up in a panic from the table, screaming and scattering in different directions. Papa yelled loudly, telling everyone, "Get the children and run outside. Now!"

Few houses remained standing as the earthquake's violent tremors shook the entire town.

Mama grabbed my sister's hand and Papa pushed us all out the door just as the walls began to crumble around us. I heard my Nana cry out as the roof caved in, trapping her and my Uncle Wira inside. Sounds like explosions surrounded us. Looking around, we realized it was the sound of houses and buildings collapsing. Papa turned back, wanting to help Nana and Wira, but Mama tugged him away, telling him we must try to escape before we were also trapped.

We looked around frantically for a way to escape. As the ground shook even harder, huge cracks opened in the earth, and trees splintered and cracked as they were torn from their roots. In the chaos, we lost track of our family as everyone panicked, scattering in different directions. Terrified figures darted here and there, desperately trying to escape the chaos. Father led us away as we ran in fear, while the earth heaved and groaned all around us.

Roads cracked open and trees were torn from their roots as the ground shifted violently.

The Causes of Earthquakes

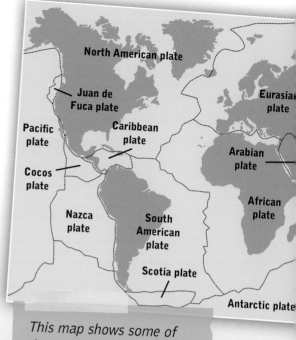

This map shows some of the world's tectonic plates.

Earthquakes most often occur along the **fault lines** where the tectonic plates collide. The surface where these plates slip across one another is known as a **fault plane**.

The San Andreas Fault

There are countless fault lines around the world. The San Andreas Fault in California is probably the best known. During the 1906 earthquake, the San Andreas Fault slipped along a plane that was 270 miles (430 km) long. The ground shifted as much as 20 feet (6 m) horizontally.

Today at this fault, the stress levels have built up so much that scientists believe a devastating earthquake with a magnitude greater than 7 is due. This means that the large number of cities located there and their dams, bridges, roads, and communications networks could be at risk.

VICTIMS OF COLLAPSED BUILDINGS

In 2008, a 7.9 magnitude earthquake struck the Sichuan province in China. More than 80 percent of the buildings in the area collapsed, destroying whole villages and towns, and burying people in rubble. Almost 375,000 people were injured, while around 90,000 lost their lives or were missing and presumed dead.

Philippine plate

Indian plate

Pacific plate

Australian plate

Natural Causes

Small earthquake tremors can sometimes be caused by meteorites. These are small, rocky bodies that enter Earth's **atmosphere** from space and collide with its surface. Other small earthquakes are the result of volcanic **eruptions**, when hot molten rock is forced violently to Earth's surface through cracks in its crust.

Human Causes

Human activities such as mining and drilling for oil or natural gas can cause earthquakes, known as **induced** earthquakes. Water pressure in cracks beneath large dams or **reservoirs** can trigger faults that are already strained. Fracking, which is a method of getting natural gas from under the ground using highly pressured water, also causes underground tremors. Underground testing of **nuclear** weapons has also been known to induce earthquakes. Although most of these types of earthquakes are small, some have caused property damage and injury to people nearby.

Poor building construction was blamed for much of the destruction and many of the deaths during the 2008 earthquake in Sichuan, China.

VIOLA'S STORY:
Escaping the Quake

At times, I felt like I was being carried along without even moving my feet as the ground shifted and rolled beneath us. Several times, I tripped as a large crack opened in the ground right in front of me, and I had to jump to safety to avoid falling in. I could hear Mama praying as we stumbled and ran. Others from the village joined us as their homes shook and swayed dangerously.

Screams pierced the air as those who did not escape in time were trapped beneath the rubble of their houses. I wanted to put my hands over my eyes and ears to keep out the frightening sights and sounds all around me.

Finally, the rumbling stopped and the ground became still. My family moved toward a large, open space where many others had gathered to avoid the falling buildings and trees. Fear and disbelief could be seen on many of the adults' faces. Families searched anxiously for other family members they had become separated from. Papa tried to use his cell phone to call my aunties and uncles who were not with us, but there was no signal.

When the Sun went down, it was really dark and everyone was frightened. There was no power. No electricity. I had never seen my town or Palu without lights. Several times, we felt shaking aftershocks that made people even more scared. They were afraid the earth would swallow us up. Luckily, these lasted only a few minutes, and no one was hurt. With nowhere to go, we had to spend the night outside in this place. Darkness wrapped around us like a tight blanket. We huddled together without food or water, waiting for the morning.

I was frightened about what had happened to our house, our family, and our village. But finally, I was so tired, I fell asleep.

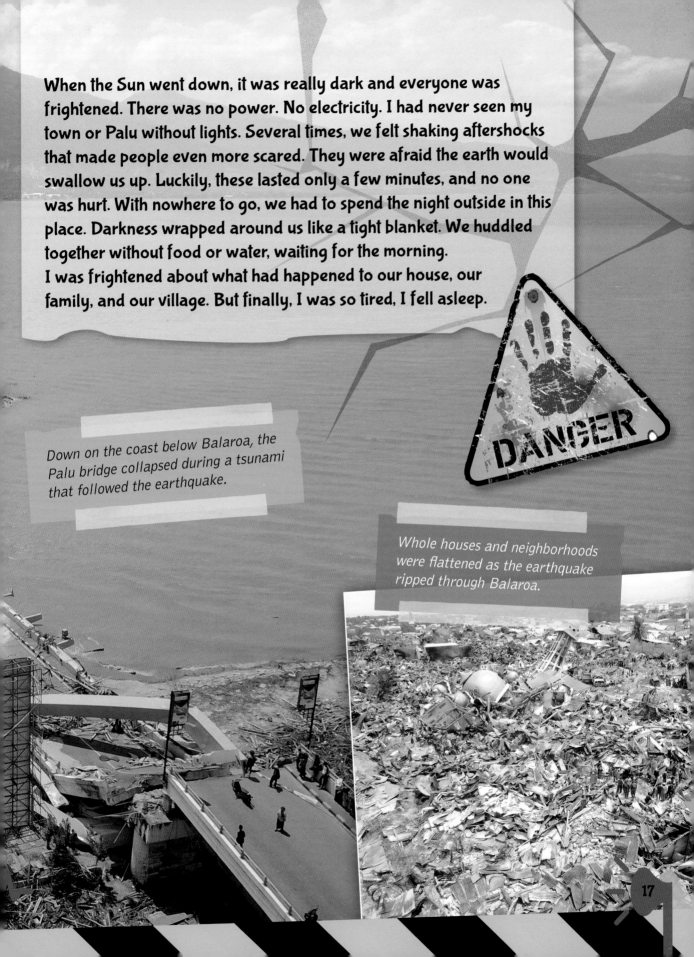

Down on the coast below Balaroa, the Palu bridge collapsed during a tsunami that followed the earthquake.

DANGER

Whole houses and neighborhoods were flattened as the earthquake ripped through Balaroa.

Where Earthquakes Happen

Each year, tens of thousands of earthquakes occur that are strong enough to be detected without the use of scientific instruments like the seismograph. Of these quakes, only about 100 are powerful enough to cause significant damage and threaten lives. Violent earthquakes, capable of causing enormous destruction and many deaths, happen only about once a year.

A Ring of Fire

More than 80 percent of all earthquake activity happens in the Circum-Pacific Belt, also known as the Ring of Fire. It is home to many densely populated regions, from New Zealand, the Philippines, and Japan to Chile, California, and Mexico. Sulawesi is located in the Java Trench on the western edge of the Ring of Fire. This is where the Eurasian, Australian, and Philippine plates meet.

New Zealand sits on the Ring of Fire. In 2011, its South Island was struck by an earthquake.

Already devastated by previous strong tropical storms, Haiti was ill-prepared to deal with the destruction of the 2010 earthquake.

The Alpide Belt

The Alpide Belt, which spans the Mediterranean region eastward through Asia, contains the African, Arabian, Indian, and Eurasian plates. Both the Alps and the Himalayan mountain ranges are in the Alpide Belt, so avalanches, mudslides, and landslides are a threat during earthquake activity.

Mid–Atlantic Ridge

The Mid-Atlantic Ridge is underwater and runs through all the world's major oceans. Numerous cross faults mark the fault, making it resemble a zipper. As they slide sideways past each other, they generate friction, which causes earthquakes. Most of this belt is far from where people live, so earthquakes there are rarely felt. However, Iceland sits on top of the Mid-Atlantic Ridge, which makes it vulnerable.

VICTIMS OF POOR INFRASTRUCTURE

In 2010, Haiti was struck by a 7.0 magnitude earthquake. More than 316,000 people were killed, and 1.5 million were left homeless. As one of the poorest countries in the world, Haiti does not have a system to deal with disasters of this magnitude. The island's **infrastructure** could not cope: electrical power systems failed, roads could not be unblocked, and communications systems failed. Many survivors suffered from shortages of medicine, food, and water because it took a long time for help to arrive.

Most at Risk

In poor countries, the homes of many people who live in big cities are constructed of mud brick, which often crumbles during an earthquake, crushing anyone inside.

As the world's population continues to grow, more people than ever are living on, or close to, earthquake zones. More cities and homes are under threat of being damaged or destroyed, and more people are at risk of being injured or killed in an earthquake.

Underdeveloped and Unsafe

Much of the world's population lives in underdeveloped countries where roads, electrical lines, communications systems, and buildings are poorly built. This puts millions of people at risk every year when an earthquake strikes. When governments cannot afford to build or maintain safe systems that allow them to deal with an earthquake's destruction, victims must wait for foreign aid and workers to arrive. Shortages of food, water, medicine, and shelter immediately following an earthquake often put many people's lives at further risk.

Living in an Earthquake Zone

The island nations of Indonesia and the Philippines are surrounded by tectonic plate boundaries, making them vulnerable to earthquakes and tsunamis. Mexico is located at the meeting point of three tectonic plates. Its capital, Mexico City, is built on the floor of a former lake. Its loose soil increases the effects of tremors, making them more destructive. Istanbul in Turkey sits on a fault line. It has 2 million buildings that are considered very unsafe.

Being Prepared

Some countries have earthquake planning programs, which help people prepare for an earthquake. For example, in the United States, California sits on the San Andreas Fault. Its roads, buildings, and bridges have been adapted to withstand moderate tremors and reduce the number of potential deaths.

VICTIMS OF FUKUSHIMA

Japan experienced a massive 9.0 magnitude earthquake in 2011. The force of the earthquake triggered a huge tsunami that engulfed coastal villages and dragged thousands of people into the sea. It also badly damaged the Fukushima Daiichi nuclear power plant, causing dangerous **radiation** to be released into the atmosphere. More than 100,000 people had to be **evacuated**, with many unable to return for years.

Some areas near the Fukushima Daiichi nuclear facility remained under evacuation for more than five years due to highly **toxic** radiation levels.

VIOLA'S STORY: After the Quake

Mama shook me awake early in the morning. Papa was desperate to return home to find Nana, Uncle Wira, and the rest of our family. What would be left of our house, I wondered? Would we be able to rebuild it, or would we have to find a new place to live?

I was hungry and thirsty because we hadn't eaten since the day before. My stomach rumbled as we made our way back home.

People's possessions lay in the rubble like they'd been thrown around in a giant's temper tantrum.

It was difficult to find our way because nothing was where it had been. Roads had heaved up. Trees lay scattered like toys on the ground. Many houses were piles of rubble, with twisted metal roofs curled up nearby. Eventually, we found our house. It was ruined. Mama sat on the ground and sobbed. There was no sign of Nana and Uncle Wira. Papa tried to use his phone again, but there was still no signal.

We learned about a nearby camp for those who'd lost their homes, so we made our way there. When we arrived, we were taken to a tent where we could stay. As more people arrived, they shared terrible stories. Everything in the city had stopped working, and no one seemed to know how to fix things. My parents were frustrated that the government wasn't doing enough to help. Some people were so angry that they broke into stores and warehouses, stealing food and water. It seemed like the world around us had gone crazy!

Papa learned that Nana and Uncle Wira had been taken to a nearby hospital. They were alive! Papa was able to find the rest of our family too. They had found a place to stay in another tent city nearby. But other people from our village were not so lucky. Some of their stories ended in tragedy when they learned that a loved one had died.

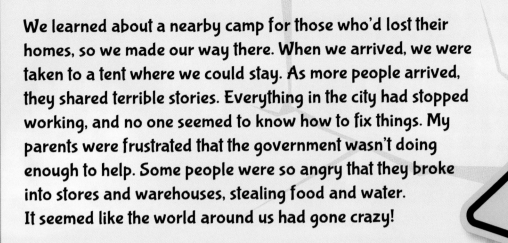

Many people posted pictures, hoping to find children who had been separated from their families during the earthquake.

DICARI ANAK HILANG

NAMA : MUH. GIBRAN , PANGGILAN : GIBRAN
UMUR : 6 TAHUN 10 BULAN
ALAMAT : JL. MITRA PERMAI KEL. KASONENA, LRG DEPAN KANTOR LURAH KABONENA
NO HP KELUARGA : 082290833677, 081341196231, 085242696769

23

How Science Can Fight Earthquakes

The science of seismology, the study of earthquakes, was developed early in the twentieth century. It began to provide scientists with answers as to where, how, and why earthquakes occur. By studying the movement of Earth's tectonic plates and its shifting fault zones, scientists are better able to **predict** a potential earthquake event.

Nature Shows the Way

For centuries, scientists and historians have noted how animals seem to behave abnormally before an earthquake. In ancient Greece, rats and snakes left the city of Helike just before an earthquake struck. People have observed fish swimming violently, chickens no longer laying eggs, and bees leaving their hives before an earthquake. Do animals feel tremors before people? Or can they detect changes in the air?

In 2010, a study found that 96 percent of male common toads in a population abandoned their **breeding site** five days before an earthquake struck L'Aquila, Italy, in 2009.

This high-rise building in Tokyo, Japan, has been made stronger by using flexible materials that bend but will not break during an earthquake.

Building Better Technology

Until now, science and technology have been limited in their ability to make earthquake predictions accurately and reliably. Most earthquake predictions still rely on past behaviors in earthquake zones to determine the likelihood of another earthquake. Scientists hope to find a more reliable way to predict future earthquakes. Until then, we must rely on better technology and materials to help us build safer cities that can withstand the forces of earthquakes.

Engineers and architects are developing new technology to make buildings, bridges, and roads more **earthquake resistant**. Better designs and "smart" building materials can help buildings absorb earthquake vibrations without collapsing. Special **alloys** and cement mixes are super elastic. They help buildings bend slightly—rather than break—during strong tremors.

VICTIMS OF BAD BUILDING

Although scientists cannot predict exactly when or where earthquakes will occur, they do know that many deaths caused by earthquakes could be avoided if people lived and worked in buildings designed to cope with these natural disasters. Around three-quarters of all deaths that take place during earthquakes are caused by buildings collapsing on people. For example, when the Sichuan earthquake that hit China in 2008 occurred, it destroyed more than 7,000 recently built schools, killing thousands of schoolchildren inside. Many of these deaths could have been avoided if the schools had been built to properly withstand earthquakes.

Protecting People

Earthquakes are inescapable natural forces that affect many people around the world. To protect people, it is essential that they are prepared and informed about what to do in the event of a powerful earthquake.

New technology allows beams to absorb some of the shock during an earthquake to prevent buildings from collapsing.

Building Better Structures

Most deaths in an earthquake event are the result of people being trapped and crushed in poorly constructed buildings. This is worse in poor countries that lack proper earthquake-resistant building materials and methods. Better construction techniques that can withstand earthquakes are needed.

Being Forewarned

Early-warning systems can give people time to move to a safer area in the event of a tsunami, landslide, or tremors. Some older buildings, bridges, and highways in earthquake zones are being changed to make them safer, but this is expensive. Poorer countries may lack money to do this, so they continue to be the most vulnerable.

Using Data to Make a Difference

The Global Earthquake Model (GEM) Foundation is an organization that works to create a world that is more resistant to the devastating effects of earthquakes. Its experts are developing tools and information to help them assess and respond to earthquake risks in countries around the world. The foundation has worked with international and regional experts in more than 150 countries to reduce earthquake risks worldwide.

Signs near the beachfront in Bali, Indonesia, mark out evacuation zones so that people know where to go.

EARTHQUAKE
EVACUATION
ZONE

TSUNAMI
EVACUATION
ZONE

FIRE
EVACUATION
ZONE

PROTECTING POTENTIAL VICTIMS

Earthquakes often strike violently and with little warning, so it is essential to be prepared and well informed about safety precautions and procedures. Have an emergency kit—including water, batteries, a first-aid kit, and a flashlight—on hand. If an earthquake strikes, people should protect themselves from falling **debris** by finding a safe shelter. They must be familiar with exits from their homes, particularly if they live in a high-rise building. People should know where emergency meeting points are, so that they can find their way there and learn what to do next. By tuning in to **media** in an emergency, people can stay informed about their best course of action.

VIOLA'S STORY:
Starting Over

As the weeks passed, we learned more about how people in our community were affected by the earthquake. More than 4,000 people died that day or were victims of mudslides, falling debris, and tsunami waves. Much of our village was completely destroyed. Although we had lost our home, we felt grateful to be alive. Some of the walls of my school had collapsed, and many of the desks and equipment were destroyed, so we went to a temporary school built near our shelter. There, our teachers encouraged us to share our earthquake stories. By talking about what had happened, we could start to understand our fears and worries.

Survivors searched the ruins for anything that they could save.

My family and I stayed in the tent city for several months while we waited for the government to provide money to help us rebuild our home. Papa returned to our old house several times to search through the debris, looking for anything he could save. But other than a few pieces of broken furniture, some tools, and a few tattered photos, there was nothing left.

Papa wanted to stay in Balaroa because of his job. And he felt it was important for our entire family to stay together and help one another. But some families decided to leave. They were afraid that the next earthquake could be even deadlier. Although we have rebuilt our home and are starting to rebuild our lives, I miss the friends who have left Balaroa.

Because Palu is located on the Ring of Fire, it is possible that another earthquake could strike at any time. The government has begun to put new rules in place to make buildings and homes safer. Our school is being built with materials and technology that will help make it stronger during an earthquake. Papa is also fighting for better warning systems to protect us from future earthquakes and tsunamis. Knowing the dangers and being prepared will help us keep safe if another earthquake strikes.

Foreign aid, including food, clean water, and medical help, arrived to help survivors.

Glossary

alloys Metals made by melting and mixing two or more metals, or a metal and another material, together

atmosphere The layer of gases that surrounds a planet

avalanches Masses of snow, ice, or rocks that slide suddenly down a mountainside

axis An imaginery line that passes through the center of Earth, from north to south

breeding site A special place where animals breed, or make babies

debris Waste or pieces of material left over from an event such as a disaster

earthquake resistant Able to withstand an earthquake

eruptions Explosions

evacuated Cleared an area of people

fault lines Fractures in the ground where an earthquake may occur

fault plane The surface where two tectonic plates slip across one another

fertile Land that is suitable for growing crops

fictional Made up, not true

induced Caused something to happen

infrastructure The basic things needed for a country or region to function properly, such as roads and schools

intensity The amount of strength or force that something has

landslides The sliding of loose earth and rock down a steep slope

magnitude The size or importance of something

media A means of communicating information, such as TV and radio

mudslides Masses of mud and other earthy material that fall down a hillside or other slope

natural disasters Disasters caused by nature, not human-made

nuclear Energy created when two atoms are split apart or joined together

predict To say when something will happen

radiation A type of dangerous and powerful energy that is produced by nuclear reactions

reservoirs Artificial lakes used to store water for use in homes

seismic Movement that is caused by an earthquake

seismic waves Energy that travels through Earth's layers as a result of an earthquake

seismograph An instrument that measures movement beneath Earth's crust

tectonic plates The pieces of Earth's crust that fit together

toxic Containing poisonous material

tsunami A series of large waves caused by an underwater earthquake

Learning More

Learn more about earthquakes and their dangers.

Books

Amson-Bradshaw, Georgia. *Earthquake Geo Facts*. Crabtree Publishing, 2018.

Furgang, Kathy. *Everything Volcanoes and Earthquakes*. National Geographic Children's Books, 2013.

Hoobler, Dorothy, and Thomas Hoobler. *What Was the San Francisco Earthquake?* Penguin Workshop, 2016.

Rose, Simon. *Earthquake Readiness*. Crabtree Publishing, 2020.

Websites

Discover the science of how earthquakes happen at:
www.ducksters.com/science/earthquakes.php

Learn about how to stay safe during an earthquake at:
www.fema.gov/media-library-data/ed51897eb6583e40ec1edb5f8fb85ee5/FEMA_FS_earthquake_508_081513.pdf

See photos from earthquakes around the world at:
kids.nationalgeographic.com/explore/science/earthquake

Explore the effects of earthquakes at:
www.usgs.gov/natural-hazards/earthquake-hazards/science/science-earthquakes?qt-science_center_objects=0#qt-science_center_objects

Index

About the Author

Linda Barghoorn has traveled to more than 60 countries. In 1992, she was visiting Egypt when a 5.8 magnitude earthquake shook the city of Cairo. Linda learned a new respect that day for the powerful forces of nature.